When The Rains Came

An Eyewitness Account Of The Deadly Tijuana Flood Of 1993

Burgin, Emmanuel, B

When the Rains Came: An Eyewitness Account of the Deadly Tijuana Flood of 1993 / by Emmanuel B. Burgin

ISBN 13: 979-8-9876721-2-9

1. Science & Math 2. Environment & Nature 3. Natural Disaster Title. When the Rains Came: An Eyewitness Account of the Deadly Tijuana Flood of 1993

Library of Congress Control Number: 000000000

Cover Design by: Jaroslav Pejoch

Illustrations Credits: Jaroslav Pejoch

Contents

Acknowledgment

"When The Rains Came" appeared first in the San Diego Weekly Reader. The author is grateful for the editorial guidance provided.

Dedication

This is a heartfelt dedication to the memory of those who perished and the bravery of those who survived.

Introduction

In January 1993, nine inches of rain fell on the San Diego/Tijuana region, seven inches more than normal for the month. Most of the rain fell within a two-week period, flooding many parts of the region. When it was over, 66 people in Tijuana had lost their lives.

Chapter One

A ntonio Romero, my brother's best friend, lost his parents on the first day of the heavy rain. In the days following the tragedy, I would come to learn from my brother Claudio the story of Antonio's loss.

On the day of the 6th, when the heavy rains began to fall, I was attending my first writer's conference at the Holiday Inn (now the ITT Sheraton Four Points Hotel) in Clairemont near Montgomery Airfield. All day, words of flood damage made the rounds among the attendees. By the evening, as the day's last lecture was about to begin in the hotel's 100-seat-theater, attendees were being informed of street closures in the North County due to the rains. Attendees who lived locally were told they would be unable to make it home that evening and advised to call loved ones and to make room arrangements at the hotel. A list of road closures was given, and some attendees left immediately to make calls while others discussed alternate routes, determined to make it home. Living only a few minutes away in Bankers Hill, I had no problem getting home. Upon entering my apartment, I turned on the late-night news for an update of the flood damage and I saw my brother's friend, although I did not recognize him at that moment, standing thigh-deep in mud with a shovel digging frantically.

From my brother, in the following days, over the phone, I heard bits and pieces of what transpired that early morning of January 6. A few weeks later, I moved to Playas de Tijuana, hoping to find some solitude while I worked on a writing project. The house was just a few blocks from my brothers'. There I met Antonio, who was now living with my brother and his family. What follows is their story told to me by my brother. The name of his friend has been changed.

Chapter Two

On January 2, a small amount of rain fell, beginning in mid-morning and continuing until late afternoon. It was nothing out of the ordinary, certainly nothing that hinted at what was to come, and by the following day the sun had broken through and all was well for the next day and a half. Then on the January 5 at about 6 p.m. a light rain began to fall. By 1 a.m. on the 6th the rain was coming down hard and the downpour would continue for the next 24 hours, and it would rain for all the next 13 days except for 1.

In Mexico, January 6, is celebrated as Dia de la Reyes (Day of the Kings). On this day, according to tradition, the three wise men arrived at the manager bearing gifts for the baby Jesus. Traditionally, it is on this day and not Christmas that Mexicans give gifts and only to the children. On this day Mexican families gather around the *rosca,* a bread formed into a wreath, *rosca* meaning wreath. Hidden in the bread, depending on the size of the gathering, are from two to eight small, plastic figurines of the baby Jesus. Piece by piece the *rosca* is cut and given out. The first to receive a figurine hosts a dinner party held on February 2, a day called Dia de la Candelaria—Day of the Candle. Others receiving figurines bring food of some kind to help the host. On the Dia de la Reyes the rains came.

Colonia Gabilondo, built like many of the neighborhoods of Tijuana, sits at the bottom of the arroyo Aguaje de la Tuna. An arroyo is a desert riverbed. Aguaje de la Tuna is one of 62 such riverbeds that run down the hills of Tijuana into the basin of the Tijuana River, which courses through the middle of the city much like the San Diego River flows through Mission Valley. Colonia Gabilondo sits about two and a half meters below the rest of the city. For many years residents of Colonia Gabilondo had suffered floods. Only two years prior the city had, at the urging of the community, finally built a flood canal beneath the streets of the neighborhood. Where the riverbed ended and the paved street Guillermo Prieto began, the city had built a concrete flood basin to the underground waterway, but the underground canal was never maintained and so when it rained, the water slowly backed up and bubbled through manholes, searching its natural level and flooding the streets. However, the residents of Colonia Gabilondo, accustomed to the flooding, had built their homes up on two- and three-foot concrete foundations and so the annual waters had been nothing more than a nuisance, until now.

Over the years, the dry riverbed had become a dumping ground. Abandoned cars, tires, appliances, bed springs, concrete slabs and garbage filled the arroyo. When the rains came, the water slowly pushed everything along until the junk became a dam. In the Colonia, after only five hours of heavy rains the water had begun coming out of the manholes. In the early morning of the Day of the Kings, the dam broke. Cars and worn-out appliances gave way and a six-foot-high wall of water rushed down the hill through the ancient arroyo.

Where the riverbed ended and the middle-class neighborhood of stucco homes began, the water swept into the streets like a train.

Chapter Three

The first home went into pieces; the wall taking the force of the water collapsing inward and then as the water rushed in, the other walls exploding outward, furniture shot out of the windows as if by cannon. The second home in the path saw the same fate. The wall of water, now searching the least resistance, rushed onward along the street, picking up cars, knocking down telephone poles. The muddy river swept up children and the men and women who chased after them. The sound was deafening, a cacophony of cars smashing into cars, refrigerators hurling against walls, couches tumbling end over end.

"It roared," one neighbor of Antonio said. "It roared and screamed."

After 60 yards, the deluge slammed into a concrete wall and a small, sloping hill and changed direction, turning with the street at a 90-degree angle. Eighty yards away, the street made a left turn, but there was no hill to redirect the surging river. In its wake was only a small brick wall and the stucco home of Antonio's parents.

Whenever the rains came, Antonio would never be far from his parents. His father was an invalid and had difficulty walking, and his mother had grown old. The waters that flooded the streets had to be watched carefully and so every half hour, Antonio ventured out onto the street to check the water level.

His brother's Kombi (VW bus) had been parked next to the house in case the water became too high, and Antonio needed to evacuate their parents. At 1:30 in the morning, Antonio called his brother and said that the water level was getting high, and he wanted to get their parents out. At 1:40 a.m., as Antonio gathered a few of his parents' belongings, the dam one mile south of them had burst.

The water rushed over the eighty yards of streets in seconds. A van, surging on the wall of water, acting as a battering ram, broke through the brick wall. The river, instead of turning, continued on through the home of Antonio's parents.

Antonio had gathered his parents in the living room and was about to step outside when the water, sounding like a train, came crashing through the front of the house and quickly filled it. The water rose to the roof. At the instant Antonio had heard the roar, he wrapped his arm beneath his mother's arm and across her chest. The water swept them into the bedroom and with his right arm he grabbed his father by the arm - who was still in the bed's corner. Within a matter of seconds, the roof came off the house, and the water swept them over the wall.

No sooner had he grabbed his father than the surging waters ripped his father from his grasp. Antonio then fought to hang onto his mother. As they were being swept along, Antonio reached out with his free hand and grabbed a mattress. Neighbors later would say they saw Antonio and his mother fighting to hold on to the mattress, some even hearing her shouts for help. The water carried them two blocks and rammed them into a wall that was under construction. Hanging onto his mother, Antonio reached out and grabbed onto a rebar with his free hand. He fought against the force of the water, holding onto his mother, knowing that if he let go she would die, yet the rush of the water was too strong. It took her from his arms. She was gone. All that was left in his grasp was the sweater she had worn.

"At 3:00 a.m. Antonio's brother called me," Claudio recalls. "He asked if I had seen Antonio or his parents. I told him that the last time I had seen Antonio was about six o'clock at the office. He then said, 'well everything has moved 'Todo está movido' I thought I had misunderstood. Maybe he had said that they had moved out of the house.

"That next morning, I had planned, after dropping a client at the airport, to meet Antonio at the hotel Ramada Inn at a political breakfast meeting. I was president of the local chapter of the PRI political party, and Antonio was the secretary. In Playas de Tijuana, it had been very quiet. Playas is a bedroom community of Tijuana and is isolated, and we had no way of knowing the devastation the rains had caused. As I tried to get across the city and the Tijuana River to get to the airport, I found it impossible. The shopping mall in the Rio Zona was flooded. It was incredible. I had never seen anything like this before in the city. So after about 35 minutes I turned on the all-news radio station, because I thought what had happened here was something serious. I couldn't get my client to the airport, so I took him to the hotel because I didn't even know if there was a way to the airport.

"The river had overflowed, as well, and with the creeks carrying everything in their wake, rocks, mud, gravel, debris strewn the streets of Tijuana. When I got to the hotel, some of my fellow political members told me they had heard that one of the hardest hit areas had been the Colonia Gabilondo. I became worried about Antonio and his family. So I headed back toward the center of the city along Boulevard Caliente, which runs along an elevated part of the city. I got as far as the gas station near Novaca [the Baja California radio station] and that was as far as anyone could go because the river that runs through Colonia Gabilondo crosses there and goes into the Tijuana river, but I had never seen anything like this. The water was just going on top of the street. And the gas station pumps had been swept away and cars had been pushed up against buildings, and the water was flowing across like a huge river with rapids, flowing over cars, bicycles, big rocks and refrigerators. It was a big muddy river.

"So, I turned around and headed back, as if I were going to Tecate, and I went up over the hill. There were rocks and huge boulders everywhere, and creeks that had never carried water were full and spilling out onto the streets. Finally, I got up the hill and went all the way around on Lomas Aguas Calientes Boulevard. I had left my house at 8:00 a.m. in the morning and by 10:30 I had made it to the home of Antonio's parents."

A drive my brother routinely made in 15 minutes, instead, took much of the morning. He had come through on the southern part of the city and as he drove he listened to the radio. The radio announcers had begun to receive distress calls from all over the city.

"A girl called and I thought I recognized the voice. I thought it was Antonio's sister. She pleaded to the public that if they had heard of Antonio or anyone in the family to please contact her. I had my cell phone, so I immediately called the number the girl had announced. It was Antonio's fiancée, Eva. She was concerned because a television broadcast from a helicopter had shown that there was nothing where Antonio's house had stood and no one could locate him.

"I became more concerned and told her not to worry, that I was a few blocks away and as soon as I found out anything I would give her a call. I also told her that I would notify my wife to see want else was going on. I drove to the top of the street about a block from Antonio's house. I saw nothing but debris and army troops and people with shovels. There was a bunch of things going on. I could see that something terrible had happened. Houses along the southern side of the street had mud to the top of the windowsills. I saw a pile of debris: boards, trees, a refrigerator and didn't realize until later that was what was left of Antonio's house. I didn't realize it because I was in shock seeing the army troops and everything else going on. I walked through the mud, over and around boulders. A woman, standing on the hill, who I recognized as a neighbor of Antonio, called out to me and said, 'Last night I saw Antonio with his mother floating by.' She said that they found Antonio about 3:30 in the morning on top of what was left of a neighbor's house in a daze. More like shock, I think.

Chapter Four

"When I got to him, I saw that he was cut and bruised from the boulders and debris. He was wearing a yellow rain jacket and pants and rubber boots. His whole family had sets of this clothing in the garage because of these constant floods. The outfits looked like those of firefighters. I didn't notice anything else but Antonio digging. He was digging next to what had been a eucalyptus tree. He was digging slowly, mechanically.

'Antonio, what happened?' I asked. He said something about it raining around 1:30 or 1:40 and that it rained hard. I still hadn't noticed that his house was not there: no car, no garage. I was focused on him. I remembered too that two years ago Antonio had dug out a drainage ditch, and, also, homes in Mexico have *pilas* [underground water tanks] and when it floods you have to dig out the mud and clean out the tanks. I saw the soldiers digging and knew something was going on, but I couldn't grasp the situation because I had never seen anything like this. I had no point of reference.

"I asked him where his mother was and he answered Red Cross had her, that a neighbor had told him that the Red Cross had taken her. I told him that we should go to the hospital and find out, and then I asked where his father was. He answered that he couldn't go until he found his father. 'Where is your father?' I asked.

'He must be underneath here,' Antonio said.

"He was digging under what was left of the tree. It was then that I looked around and realized that there was no house, no nothing. It finally hit me. I finally grasp what happened. Antonio wasn't crying or showing any emotion. I tried to get him to come with me, but he didn't want to leave. I called my wife on the phone to let her know what was going on, but she already knew. The TV crews were already there, and she could see Antonio with the shovel. A few minutes before I had arrived, a reporter had interviewed Antonio. I then called Antonio's fiancée, but he didn't want to talk to her. He finally got on the phone and, as he began to explain to her what had happened, he began to cry. He told her that within seconds his house, everything was gone.

"I told Eva that I couldn't believe what I was seeing, because I remembered the house — even the concrete driveway was gone. The flow of the river had been so strong it took everything, even the concrete foundation. Nothing, nothing was left. It was just one muddy empty lot," Claudio said.

Reports began coming in on the dead. An elderly man, a neighbor who had known Antonio's father and had seen Antonio swept away, suffered a massive heart attack. Another man, who had tried to come to the aid of the heart attack victim, had been electrocuted when he grabbed onto the wrought-iron fence which had fallen wires lying across it. The Red Cross was taking their bodies out of the yard as Claudio stood by Antonio. The electrocution led to the soldiers to evacuate everyone. The army had now taken control.

"I told Antonio that we had to go, but he said he couldn't leave. He refused. I tried to persuade him by telling him he needed to contact his brothers and sisters. Antonio had a brother in Texas and a sister in Missouri, and another sister in Chula Vista. A neighbor told me that Antonio also had an aunt, who lived in Colonia Cacho, a neighborhood nearby.

"It was very cold, and I was dressed in a suit and trench coat. If Antonio was not going to leave, I needed to get more adequate clothing. I told him I would return to help him," Claudio said.

My brother changed and by the time he returned, the army had removed a great deal of mud from the area of the street Antonio had been digging, yet they had been unable to locate the body of his father. Soldiers in rain gear and driving small bulldozers from the Army Corp of Engineers filled the street. The big earthmovers had not been brought in because many people were believed to still be in the mud. They were still trying to account for the people in the nearby homes. Red Cross, the fire department, and the police were all there.

According to Claudio, when the Mayor of Tijuana, Héctor Osuna, who had taken office three weeks prior to the flooding, saw the soldiers, he reacted in anger. Osuna was of the PAN party, which for the first time had won not only the mayoralty but the governorship of Baja as well. He had not declared Tijuana a disaster zone, and since he had not, the federal government, which was the PRI party, could not come in and assist.

"He needed the help of the federal government, and he didn't want that. Unfortunately, he was thinking of his political problem rather than of helping the people. The mayor said that everything was under control when that was false: there were parts of the city no one knew anything about because these areas were not accessible.

"I returned in an hour and a half and brought with me some sandwiches and Gatorade, but Antonio refused to eat, and he still wouldn't speak. I began to speak with the radio stations on my cell phone. Anyone who knew Antonio or me knew we were there and so through the stations, I received calls. While I talked on the phone, I stood next to Antonio as he continued to dig in the same place next to the tree along with the soldiers who had not yet been called off.

"When I was at my office, I had told my employees that Antonio had nothing: his furniture gone, his clothing, his parents missing, and I was still under the belief that his mother was fine because of what the neighbor had told him. But upon my return around 2:30 p.m. I had found that information to be concocted, a report just to keep Antonio going. Antonio thought his mother was okay but knew his father was gone. That much he knew. So I told all my employees to call the hospitals and Red Cross to see if they could locate his mother and to see if they could locate his other family members.

"When I returned, I met Antonio's brother. He said Antonio hadn't eaten anything and that he had been digging since about 3:30 in the morning, almost 12 hours ago. In all that time Antonio had refused food and fluids, though he had accepted a couple of cigarettes.

"At about 5:30 p.m. it started to get dark and the soldiers said there was nothing more they could do that evening but that they would start again in the morning. The soldiers had been trying to get lights and to get everything organized, but it was too dangerous. And with fallen cables and not knowing everything that might be in the mud and debris, they could not have anyone walking around there. So the area, as well as the whole section of town, was cordoned off," Claudio said.

As the soldiers and police busied themselves with setting up barricades, people still moved about the damaged houses, carrying belongings out. Homes on the southern side of the street had mud spilling out of their windows. Of the three homes on the eastern side that had taken the brunt of the force, only the home on the corner was upright. It sat crooked and off its foundation and its basement filled with mud. The other house, like Antonio's, had been obliterated.

Antonio's dark green stucco home with the porch and swing, the brick fence with wrought iron, and the concrete driveway landscaped with the roses had been wiped off the face of the earth. The big, gnarled, old eucalyptus tree was nothing more than a stump. The neighbors stucco white house with its porch and red roof, also, gone.

In the three hours since Claudio had returned, enough debris and mud had been cleared so that it was possible to walk a couple of blocks, although with difficulty.

"The mud was still deep, maybe up to mid-calf and people were looking for bodies and they were saying that because some of the manhole covers were off that, perhaps, some of the people missing had been sucked under and into the river that ran underneath. I was hoping that was not the way they were going to find Antonio's father," Claudio said.

A half an inch of rain fell that afternoon with the temperature hovering near 50 degrees. As darkness enveloped the street, Antonio still had refused food and fluids.

'Antonio, where are you going to sleep tonight?' I asked. He answered that his brother was here with him now and his aunt was a few blocks away. He guessed he was going with them. I didn't have the aunt's phone number and I knew I was going to lose contact with him, so I told Antonio that I would meet him here tomorrow as soon as it got light."

Antonio agreed and Claudio left him a half a pack of cigarettes and a lighter. As he made his way through the mud and had gone about half a block, he heard Antonio call out to him.

'Claudio.'

'¿Qué pasó Antonio?"

'¿ Qué va a ser de mi?' (What is to become of me?) Antonio said with a hand outstretched to one side.

'Don't worry about it. As long as I am here, you are going to be okay. We will find your parents. You're not alone. There are no words to express what you are going through, but don't worry. As long as I am here, you have nothing to worry about,' I answered."

Antonio turned around and walked away and Claudio then turned and walked in the opposite direction.

"It was devastating for me to talk to someone who had lost everything, lost his parents, his home, pictures, everything that had to do with his life, his family albums, his stereo, nothing, nothing left. He didn't even have a pair of sneakers," Claudio said.

Chapter Five

On the following morning of the 8th at 8:30, the local community was out in force to help Antonio. He had lived there all his life and his neighbors were rallying around him. Somehow it seemed that the neighborhood could not rest until his father was found.

Standing some 50 yards down the street from the eucalyptus at the bottom of the hill where the digging was now taking place, it dawned on Claudio that all the debris around him was what was left of Antonio's home. Boards and chunks of stucco had been carried there. That morning they had begun digging in earnest, thinking, if the debris had carried there, perhaps, then his father was also there.

On this morning though, there would be no help from the city or the police. The city-wide disaster had stretched their forces beyond capacity. In search of help, Claudio called his office at PRI and told his associates that one of their members needed their assistance.

"I told them what happened, that his father was here, and to check hospitals for the mother. Maybe she was unconscious. I told them that I was going to go to city hall to ask for help, because the city had pulled out the machinery. What had happened was that the disaster had been too great for the city.

"I went to city hall and stated that these people needed help. But at city hall they told me of the magnitude of the problem that there were still parts of the city they had yet to get into. I explained the situation to the mayor's assistant and was told only that they wished they could help.

"I asked why hadn't the mayor called the city a disaster zone so we can get the federal government in here. They can send in planes with equipment. They can send in helicopters. He was handling it as a political party issue. So the mayor couldn't do anything because he didn't have the equipment, and without the soldiers they couldn't send people to help.

"The mayor's assistant said the Americans had offered to send people down but that the mayor wouldn't accept that either, saying that he would have to get federal authorization. The mayor, because of politics, left the people to fend for themselves," Claudio said.

Obtaining no help at city hall, Claudio went over to the party offices. There he was told that they had members who were bringing in small Caterpillars and that they would send one over that afternoon or tomorrow morning as soon as they located one. So that afternoon everything was done by hand and shovel. When he returned around 1 p.m., he saw that Antonio was not by his house or down by the debris. He had moved across the street to a house with part of its concrete wall still upright. The water had gone around this wall and had gutted the house, but between the wall and the house, pictures and documents from Antonio's home had been found.

"Antonio had placed a board from the wall to the window of this house and he was lying on his stomach sifting through the muddy water and pulling documents. Workers figured if they could pump this water out, maybe they would find his father's body," Claudio said.

For the next two days rain fell intermittently as Antonio worked in this area, and, as they found pictures and documents, he would dry them out. It was all he had left of his parents.

"Antonio still was not talking. His brother from Texas and sister from Missouri had arrived. When his sister made it to the site, they called him and he went over to see her, but he was in a sort of a daze. She got him inside a neighbor's house and it was then that I thought that he might break down, but he didn't. She was hugging him but his arms were down at his side, and then he went back to digging. They left him alone and then his brother and sister got shovels. The sister from Chula Vista was there too along with her neighbors, and they were all digging. In some areas there were cars buried under the mud and firemen were checking them for bodies," Claudio said.

During the night it rained hard and in the morning it drizzled. News reports were saying some 30,000 people were homeless and the radio stations were swamped with calls from people saying that they had not seen any firemen, that no one from the city had come out to help them.

"They were all clamoring," Claudio said. "So that night, when I got back to the site, I got on the radio station and said that everyone should send a telegram to the president of Mexico. I said tell him that we need the federal government in here to help us, because the impression the mayor was giving out, through his bulletins, was that everything was under control and that they were providing homes and places where these people could sleep and that they were giving them food when, in fact, it wasn't so. The mayor was even appearing in commercials saying everything was under control."

By the third day Antonio's mother hadn't been found in any hospital. Claudio had heard 18 bodies lay unclaimed in the city morgues and so on the morning of the 9th, he began searching the morgues.

Back at the site, dozens of people continued to dig. A water truck was brought in and began pumping water out of the house.

"Without the city's help it was just neighbors doing the digging. Everyone had gotten pretty well organized. Everyone was focused on Antonio. In the entire city he had been the most affected and his neighbors were trying to do everything they could to help him. Throughout all this he had continued to refuse to drink or eat. I was very concerned for his health, so I called a friend Dr. Avila Gil.

"I told the Doctor the problem, and I suggested injecting Antonio with a tranquilizer to put him to sleep, but the doctor said that would be the worst thing to do, that Antonio was in shock. He said when Antonio's body tired he would eventually collapse and to let him be, but that he would give him a shot for tetanus because of the scrapes and the filthy water," Claudio said.

When told he needed a shot, Antonio refused, explaining he didn't want a shot that others had tried to give him tranquilizer shots but that he didn't need it. He said he couldn't do that until he found his father. But Dr. Avila Gil reasoned with Antonio, telling him that if it were important for him to find his father he would have to keep his health, and in order to keep his health he needed a tetanus shot and fluids.

"He drank some Gatorade, took some vitamins, and got his shot. After that he began to take in fluids," Claudio said.

As they slowly pumped water out of the house over the next two days, many of Antonio's and his family's' items were found.

"His toolbox, his clothing, a suit, shoes, the chest of drawers that had all the house documents were found, even the TV and what was left of the stereo, and then his mother's wedding picture and other family pictures were found and this began to make him feel better. He dried everything on wooden crates, barrels, hangers that had been scavenged from the mud," Claudio said.

Finally, on January 12, six days after the flooding and loss of life, the mayor succumbed to the pressure of the people and declared a state of emergency. This was also the only day that it did not rain during the two weeks. Within hours, military planes loaded with equipment, blankets and emergency kitchens landed and military personnel began to deploy throughout the city. At 1:00 p.m. the federal government flew in Carlos Rojas, the head of a public improvement program. Arriving at the airport under heavy security, Rojas held a quick news conference. As he began to speak, one of Antonio's sisters, under the guise of a news reporter, ran up to Rojas and began to tell him of her family's ordeal. She had obtained credentials from a sympathetic media fed up with the mayor. The governor, who was of the same party as the mayor and had supported his decisions, stood by quietly, his face turning red with anger.

She told Rojas that the Americans had offered dogs to search for bodies, that she had lost all hope of finding her parents. There were many unaccounted and two of them were her parents', she said. The mayor and governor looked at each other, wondering how she had gotten into the secured area.

The mayor, trying to save face, responded that only the federal government could authorize help from another country. Rojas told her and the press 'if that was what she needed, then that was what she would get.' (Unbeknownst to her and Rojas, search and rescue dogs were already in Tijuana).

Rojos said that he had been briefed and understood what was going on and that the federal government was here to help and would do everything possible to support the people. The president, Rojas stated, had received such a considerable amount of telegrams from the people of Tijuana that he had sent the Secretary of Defense to Tijuana to observe and report back. The president felt his administration had not been given an accurate picture of what had occurred in the city, Rojas remarked. When his beliefs were substantiated that things were not in control, he had been ready to send in the army to support the people, whether the mayor declared a disaster zone or not.

By the end of Rojas' new conference, the army was back helping the community.

"This was something the government always had for these emergencies, but the city wasn't getting it because the mayor nor the governor would ask for it," Claudio said.

A few hours later, President Salinas arrived.

"It was a touchy situation politically because the city government was led by PAN and the federal government by PRI. From the airport, after a brief news conference, President Salinas went directly to where Antonio was still searching for his father. Of course, once the local politicians knew the President was coming, they had all kinds of machines there, even a firetruck. The Red Cross had a kitchen there giving out pizza and tacos. They had so much machinery they were getting in each other's way. "Workers and soldiers had cleaned up the streets. The cars that had been buried had been carted away. You could actually walk down the street. It was still muddy because every night it continued to rain. But neither the mayor nor the governor wanted to get his feet dirty or muddy. They tried to show the president everything from the top of the hill above the street, but the president got out of his car and walked down the hill through the mud to where Antonio's house had been and asked for Antonio by name.

"He asked Antonio how he was. Antonio responded that they had not been able to get any help. He told the President that he had been to the mayor's office himself and that even though they had said it was too much for the city to handle, they would not call the city a disaster area. The president told Antonio, 'You are not alone and we are going to find your parents.'

"The president took off and went around the corner with the press and, Secretary of Social Development Luis Donaldo Colosio Murrieta [a future presidential candidate, who was assassinated in Tijuana in 1994] stayed behind. We knew Colosio from the party and I walked with him and Antonio. Colosio said, "This is something terrible. It's unfortunate what has happened politically. They are handling this as if it were political. It's not political. The first thing is to help the people and then worry about who receives the credit. So sad that it has taken this long for us to come here, but it is the politics. We just can't come in. It is a sovereign state," Claudio said, recalling Colosio's words.

They walked with Colosio back to his car, where the president continued to speak to the press. After a few minutes, President Salinas and Colosio were in their cars and heading to the airport.

Chapter Six

The 13th was Antonio's birthday. It had been seven days since his parents had been swept away and Antonio still would not talk.

"All he did was look for papers. I was still trying to find a way to reach him. He always liked my kids. He would come over to the house and help Erik, my youngest son, fix up his bike. Antonio is Erik's godfather. I thought maybe if he saw Erik, it would cheer him up. This condition had been going on for a week. So my wife and kids drove down there. They hadn't been there. They had only seen it on TV.

"I took the kids to Antonio. My wife stayed back at the corner because it was still muddy. They went up to Antonio and wished him a happy birthday. He was digging, of course. He looked up to acknowledge them, but he wouldn't talk, then he returned to digging. Erik said '*Feliz* cumpleaños.' [Happy birthday], but he kept on digging. I told Antonio that I had to go to the bank to make some business deposits, but we would be right back. They brought him some Gatorade and some *tortas* and left them on top of a barrel. He nodded but he wouldn't talk.

"No sooner was I at the top of the hill and in the car driving away that I here on the radio Antonio's name mentioned. By that time, ever since the president had talked to him, Antonio had become headlines throughout the nation. The broadcaster said that 24 hours after the president had authorized the dogs to come in and search that they had located his mother.' I said, 'Whoa they found her.' I tried to get to where they had located her, but the streets were blocked off."

One block south of Boulevard Agua Caliente at the same intersection Claudio had been unable to navigate on the first day of the flood, where the water looked like rapids as it rushed across the boulevard, they found her body inside the cafeteria of the Pinturas Corona paint factory.

Factory workers, who had finally been allowed to begin the cleanup of their plant, discovered the bruised body as it lay uncovered beneath a table.

Failing to get to the body, Claudio returned to Antonio, again having to take the roundabout way, approaching from the south. By the time he reached the street there were so many news reporters and vans that the street was blocked. He took another street, the same one the president had taken. From atop the hill Claudio made his way down to where the house had been. Soldiers had cleared all debris and were now using a bulldozer to scoop the mud. At the moment Claudio arrived, the bulldozer lifted a load of mud and in the mud was the body of Antonio's father.

"Luckily my sons had stayed in the car. As I ran down the street, I could see Antonio and his family gathering around the body. Within 15 minutes they had found both bodies of his parents on his birthday.

"People were saying it was such a sad thing to find his parents on his birthday, but I said, 'No, think of it as your birthday gift from them. They wanted you to be at ease, Antonio.' And that was the way it was. The very next day was the funeral and the skies opened again and it rained terribly hard. And the Zona del Rio once again flooded.

"After the funeral I took Antonio's family in the company van that sat 16. After they buried his parents, I gave Antonio a hug. Everyone was in hysterics and crying. Antonio was still and quiet. He spoke very little, but he did say to me that he did not want to ride with his family. One of his aunts had a house in another part of town and the family needed to get together. So that is where he wanted to go. He asked if I would give him a ride and then he called for his nephew to come along. Antonio always liked children. He was good with kids and so Antonio, his fiancée, and nephew rode back with me and my family. I said you need to eat now, and he said, 'Yeah I can eat now'. We took him to Denny's," Claudio said.

In the following days, a collection of money and clothing was given to Antonio by the Travel Agency Associations, of which Claudio was a member. The PRI party and the church also donated money and clothing. Claudio went to city hall and obtained a new birth certificate and passport for Antonio, which, according to Claudio, was issued without the documentation usually required.

"Even the American consulate did not ask for anything when Antonio's visa was reissued. Antonio was pretty well known in the city by then."

Antonio stayed at his aunt's house for about a month. But Claudio, knowing his sisters and brothers would have to return to their lives in the states, invited Antonio to live at his home. For a month he lived in the bedroom of my brother's oldest son, and when the small apartment in the back of the house became available, Antonio moved in and returned to work. But Antonio soon found himself involved with courts battles. He and other residents of

Colonia Gabilondo were having to prove ownership to qualify for emergency funds and they were also suing the city for lack of responsibility in maintaining the drains and sewers properly. The legalities took so long and Antonio became so involved, eventually becoming their spokesperson, that he had to take a leave of absence from the travel agency. The lawsuits took over a year and once that was done he married his fiancée, Eva, and they now live in Playas. Soon after the court settlements, he began working for the Ministry of Urban Development, helping poor neighborhoods that need sewers, drains and paved streets.

Five years later, I am again living in Playas and working on a writing project. Sometimes, as I drive around Tijuana and see the newly paved streets, wide gutters and drains, even the newly planted trees, I am reminded that it has all come at such a price that, for most of us, is unimaginable.

AFTERWORD

The tragic flood of '93 could have been averted had the political parties prioritized the people they are supposed to represent over their own agendas.

PRI lost the election to PAN July 1992 and Mayor Hector Osuna Jaime, representing the PAN Party, officially assumed office on November 1st. In a sweeping move, he dismissed all previous city officials and replaced them with inexperienced members from his own party. Instead of considering the community, Mayor Osuna and his party PAN were concerned solely with promoting their own party members. Regrettably, this had tragic consequences.

As per usual practice, the drainage canals and sand traps, specifically designed to safeguard the city's drainage network from debris, were subject to annual cleaning prior to the onset of the rainy season.

The inexperience of Mayor Osuna's appointed officials led, deplorably, to a lack of attention towards maintaining and clearing the drainage canals, as well as the substantial rubbish-filled sand traps. This act of negligence resulted in the overflow and subsequent loss of sixty-six lives. Among them were the parents of my good friend, Antonio.

These political parties, now eight in Mexico, think only of themselves. However, this issue does not limit itself solely to Mexico. Parties in Russia, Cuba, Venezuela, Nicaragua, China, and yes in the United States think and work for their cronies and not the people. This is a noteworthy point for contemplation.

The historical journey of Tijuana continues as it grows bigger and stronger. Life continues. My friend Antonio recovered, married, and started a family.

Hopefully, people will not forget this tragedy, and the parties involved will learn to serve their constituents and learn the power of cooperation. That is my hope. That is my wish.

Sincerely,

Claudio Burgin

BONUS

Identity, Racism, and Violence:

As a Mexican-American, the terms "identity," "racism," and "violence" have been integral parts of my vocabulary for as long as I can remember. These words not only shape my personal experiences, but also influence the broader narrative of what it means to navigate life in a society where my heritage is often both celebrated and marginalized.

The history of Mexican-Americans is replete with racism and violence, and the current socio-political climate, with its heightened racial tensions and divisive rhetoric, exacerbates this violence.

Despite these challenges, the Mexican-American community continues to show resilience and strength. Our identity is a source of pride and a reminder of the rich cultural heritage that defines us. Through activism, education, and solidarity, we work to combat racism and advocate for social justice. Organizations and movements that focus on civil rights, immigration reform, and community empowerment play a crucial role in this fight. Additionally, representation in media, politics, and other spheres is essential in changing narratives and challenging stereotypes.

Understanding and addressing these issues is crucial not only for the Mexican-American community but for the broader goal of achieving equality and justice for all marginalized groups. Through collective action and a commitment to equity, we can create a society that celebrates diversity and provides everyone with the opportunity to thrive.

The three pieces plus the short story "The Sugar Brown Store" included in this special print edition demonstrate my exploration of the Mexican-American experience.

Emmanuel Burgin
PEN America Member

Villaseñor tells magical tales at book-signing

El Sol de San Diego-November 24, 1994

By Emmanuel Burgin

Victor Villaseñor, the critically acclaimed and best-selling author of "Rain of Gold" and "Macho!" read from and signed his latest book, "Walking Stars", on Nov. 12 at Esmerelda Book Store in Del Mar.

"Walking Stars," a collection of inspirational tales written for children ages through 12, is gathered from the author's family tradition of oral storytelling. The book's intent is to spark the imagination and build self-confidence and a sense of wonder in children.

Villaseñor spoke of the importance of allowing a child to dream, to paint, to sing, and to be in touch with the power and magic of the spirit.

"There is another whole world of reality," he stated.

He recalled the moment, at age 35, when his "world exploded." After struggling for many years with dyslexia, he had become a successful writer and wanted to write his big book, which would become "Rain of Gold." However, he could not begin it because he had become Anglicized and had lost his ability to believe the (magical) stories of his mother and father.

But when his father's story of the extraordinary serpent, which stood as tall as a man on horseback, was corroborated by a San Diego Zoo staff member (the Bushmaster snake grew to 14 feet and could rise up to half its length, which made it quite dangerous, and therefore, near extinct), it made him reassess his doubts.

"Maybe all my parents' stories were true. Maybe it's just me and my reality that isn't able to understand it," he said.

With a new willingness to be open, he decided to return to Mexico. It was at a dinner he hosted that the true catharsis happened, not only gave the author his understanding of the power of a life truly rooted to nature and tradition, but which also gave birth to the title of his book "Walking Stars.

"I looked out the window and could barely make out the black jagged mountain from the night sky. And then I saw the stars beginning to move downward. I asked, 'what was that?' and was told that those were the guests I had invited to dinner.

They were walking down the mountain with torches. 'I thought they were walking stars,' I said. I mistook them for stars. But a man said that was what they were. I said I understood. They're people, but for a moment I thought they were stars walking down the mountain: Again, somewhat impatiently, the man told me that that's what they were, and hadn't my mother told me that's what we all were? We were all walking stars coming down from heaven. We were all visiting earth here.

"Then I realized at that second that that's what the man meant, and that when I was a little boy and my mother had put me to bed (telling me that I was an angel), she had meant literally that I was an angel, that every child born was an angel, and that we came from the heavens.

"When I realized that they truly meant this, that it was not symbolism, my whole world exploded, and I began to understand that these people who were truly rooted to the sky, to the night, to the sun and the moon, that like a dog that can smell and hear what we can't, these people were rooted to the daily miracle of life in a way that we had lost."

Thus, "Walking Stars' is the author's call not only to the children but to the children in all of us, that we must never forget our traditions and culture, our abilities to believe in the power of nature's magic, and in the power of ourselves.

Unless you are proud of your roots, you are dead. Each of us must become rich in our culture before we join main-stream, or we are hollow bones, we are hollow lost people."

A Chicano in Prague

Oppression: Whether communist, fascist or economic, the power to degrade people's lives speaks the same language.

Los Angeles Times

August 30, 1996|EMMANUEL BURGIN |

PRAGUE — Am I the first Chicano in Prague? Rudolfo Anaya, the esteemed Chicano writer, can lay claim to being the first Chicano to travel to China. It's from "A Chicano in China," his journal of that journey, that I derive my title.

Although it would be an accomplishment, something to tell the grandkids, I doubt that I am the first Chicano in Prague; we travel everywhere now, borders never having grasped our imagination, akin, perhaps, to the Native Americans' inability to conceptualize the owning of Mother Earth.

My sojourn here to live among the people has been twofold: that I may begin to understand the effects an oppressive communist regime has had on its people and to witness the struggle to Westernize in the face of the tidal wave called the global economy.

In so doing, I hope to better understand the intolerant attitudes that sometimes rear their ugly heads in my country. In California, oppression of rights, the plague of the downtrodden, has joined with economic struggle and taken a place at the kitchen table of the immigrant and working poor. The battle to retain rights granted by the Constitution, compounded by the historical economic struggle that all immigrants have known has created a crucible in which something volatile is brewing.

I was greatly alarmed and affected by Proposition 187. Not so much by the politics and the rhetoric of the politicians–after all, politics is a dirty business–but by the public's prevalent eagerness and acceptance of this mean-spirited rhetoric.

There is always a deeper meaning behind the action. A wildfire needs grass, shrubs and trees to consume in order to move forward. I am alarmed at the deeper meaning of this acceptance: the insensitivity of a friend, the latent racism of a kindly neighbor.

Language that is designed to separate and abuse and spread fear, in other words, oppress, eventually settle comfortably into the laps of those whose ideas of what a good society should be are, those who are, in the extreme, racist and, to say the least, not very understanding of the multiethnic society that we are.

Dialogue of oppression can be wrapped in many colorful packages (economic stability, rights of citizens, unfair tax burdens, crime) but it is still the language of oppression, words that fan the flames of frustration, anger, hate and racism.

It is good to remember the words of Gyorgy Konrad, a leading Hungarian writer who as a child barely escaped Auschwitz, then the Arrow Cross, the Hungarian Fascist party that wanted to shoot him and dump his body into the Danube.

"Intimidating or constraining or killing one's fellow man for belonging to this or that group has become inimical to Europeans, despite their long history of racial, national and class hatred, or rather because they have learned from their history and finally realize that discrimination leads to murder."

When our representatives try to pass legislation that will divide the people and discriminate against a segment of the population, we are in dangerous territory.

Escaping The Mean Streets

Los Angeles Times Op-ed Dec. 19, 2012 [Emmanuel Burgin]

Growing up in Compton and later in Lynwood during the 1960s and '70s, I learned way more about violence than a kid should know.

I lost a close childhood friend to a bullet, and I knew at least a dozen kids at school who never made it to adulthood. Most of them died from gunshots. I remember going to high school on Monday mornings and wondering who hadn't made it through the weekend.

There were a lot of guns around. The gang members and criminals had them, and some law-abiding people had them too. My father, who grew up hunting in Oklahoma, slept with a gun under his pillow for protection after he moved to Compton. But that wasn't what most of the guns in my neighborhood got used for.

I have tried to explain to friends what it was like — and how I'm sure it still is — walking the streets of Compton and Lynwood; I always had to be aware of my surroundings. I focused on who was coming toward me on either side of the street and looked over my shoulder frequently. I evaluated every passing car.

I didn't learn until later in life how it felt to just go for a walk. No one just went for a walk in my neighborhood in those days.

In college, on a football scholarship, I fell in love with rugby. I played the game from the age of 18 until I was 33, playing the last 10 years with the Belmont Shore Rugby Club. In my last year, our coach was asked to take another team on a tour abroad, and he invited me to tag along. I decided to treat myself and go. It would be my first time outside the country other than a few road trips to Tijuana with college buddies.

As an inner-city kid, I was in awe of historic Europe; I saw the Louvre in Paris, walked past the Manneken Pis in Brussels, and sipped a beer in a public square in Luxembourg. And then we got to Amsterdam.

In the Dutch capital, our coach called us together in the lobby of the hotel to offer some last-minute advice before we headed out to see the sights. Amsterdam, he cautioned, was the wildest and most dangerous city in Europe, so we needed to be careful. We shouldn't go out walking around by ourselves, staying in groups of at least two or three.

His concern for his players was genuine, and he went on and on about our safety. Meanwhile, my teammates and I were eager to hit the streets: There were beers to drink, songs to sing and girls to chase. We listened politely to his warnings until finally I raised my hand.

"Coach, are there any guns in this city?" I asked.

"No guns," he replied. "Handguns are outlawed in most of Europe. It's illegal to own one."

"A city without guns?" I asked, getting to my feet. "I'm from Compton, coach. A knife or a bat I can outrun. I'm outta here."

Everyone laughed and the tension lifted. I headed out the door and onto the streets of Amsterdam, and for the first time in my life, I felt free.

The Sugar Brown Store

Violence has its energy, a man-made tornado. Growing up in south central Los Angeles, Compton to be exact. I had seen my fair share of it. Acutely aware of that moment before the violence, a stillness of the air, violence's companion. Its aftermath, a deadness of mind and an emptiness of soul.

The concurrence of brutality and beauty visited upon a little part of the world inspired the Sugar Brown Store.

A Short Story

by Emmanuel Burgin

The bullet that killed Sweet Jeffrey in front of the Sugar Brown Store whizzed through the currents at a velocity of nine hundred feet per second. The bullet upon entering

Jeffrey's chest was true and precise, creating a hole the size of a dime, but upon exiting his back the tumbling bullet exploded, leaving a hole six inches in diameter, depositing his right lung and scapula against the brownstone wall.

With his legs extended in front of him, Jeffery sat slumped to one side, coming to rest gently on his left temple. His left arm was twisted and pinned behind him against the brownstone wall, while his right arm was stretched out in front of him, the hand splayed. What enraptured those that gathered to see Jeffrey in his last moments on earth was how he had struck such a beautiful, poetic pose.

A shiny European suit of dark green draped his body, underneath he had worn a black silk shirt and on his feet were black Italian loafers. A gold wedding band and a pinky ring of black onyx and a gold crucifix, which had been pushed up and out of his shirt during the explosion that occurred in his body and now dangled near his chin, were his only jewelry. It was no surprise to some, "Leave it to sweet Jeffrey," that he would be the one to make such a gallant, courageous last defiance of death on the streets, because by capturing death in such a way Jeffrey was to be remembered always. For nearby was a photographer from Life Magazine who was shooting an expose of the city for "A Day in the Life."

Across town, a small young girl lost her purple balloon and instantly began to cry at the loss of the object of her attention until the movement of flight captured her imagination. She stood on the lawn of her friend's house in her pink party dress, which had been bought for her friend's birthday party, and gazed upward as the balloon maneuvered its way up into the clouds, and as her expression changed from sharp pained features into soft wonderment, this too was captured for posterity by a mother, who sent the photograph to her favorite local news show.

Carlos Terrano ran with a gun in the shadows through garbage can strewn alleys, over chain-link fences, across vacant lots, and past churches of all denominations. In his mind, he knew what he had done and repeated "oh my God" until it became a chant. And the chant became a way for Carlos to stabilize his breathing as he ran; it was a technique he had learned while on the Lincoln High cross-country team. But even with the chant, it had become hard to breathe. He had now been running for twenty minutes and calculated that he had run at least three miles. And he smiled at this calculation, and if not for the stinging in his lungs would have laughed, laughed at how little ground he had covered, at how little ground he, a star cross-country man, had covered in such a time of danger. And he tried to rationalize this by taking into consideration that he was not running the 10,000-meter race in the Dominguez Hills but, rather, an obstacle course through East L.A... But as soon as he had rationalized it, the sirens could be heard again; because as long as he kept his mind clear and concentrated on his breathing and steps, the sirens never reached his ears, but he could never concentrate for long, because of the image he carried in his mind, and whenever he slipped back into the real world, the sirens would fill his head. And now the sirens were closer than ever.

After Jeffrey's body had been taken away, after the detectives had taken their notes, after the children had mockingly lain where Jeffrey had lain, –acting out another gangster's last moments, passing the name of Sweet Jeffrey on to the ledger of street lore–the Sugar Brown store opened for business. And Carolyn, a young mother of three young boys, of which one had contorted his body to fit the outline that had marked where Jeffrey had fallen, bought a carton of milk, a dozen eggs, and a stick of butter, and with the change from the five-dollar bill bought the boys each ten cents worth of hard candy.

The little girl in the pink dress was put into the shiny new green sedan. She tugged at her seat belt, which her mother had put on too tight, and barely able to see over the rear passenger door, waved at her friend, who stood waving back with a purple balloon in her hand. And the little girl could not understand how her friend had caught the balloon, when it had been so high up in the air, so high up that not even her mother could reach it. Then the car began to move, and the little girl put her head back against the seat and felt the motion o the car until she fell asleep.

When Carlos heard the words to stop, he honestly thought that he would. He had told himself as much. But, instead, he ran on, because now he could see the end of the alley and freedom ahead of him—the tape at the finish line. With the words from the officer in his ears, he began to run in a zigzag motion, tipping trash cans behind him as he went. He heard the footsteps behind him and made quick calculations, which told him that after sixty yards the footsteps would slow to the point where they would no longer close the gap, and he would, at this rate, make it o the light at the end of the alley. He turned to look over his shoulder to confirm his calculations, but it was in this moment that a sudden flash went off inside his head, followed by a sudden warmth that spread along the side of his face like maple syrup. And then he was on the ground, sliding and skidding through rotten lettuce and wooden crates and bits and pieces of fruit that he could not recognize. He was being pushed forward by a force so great that he felt at any moment it would give him the momentum needed to regain his feet. But as hard as he tried to get to his feet, he could only crawl on his belly through the garbage, and as hard as he tried, he could not lift his head from the smell, so he closed his eyes as he so often did in races, when he was trying to run through the pain, and he held his breath because now the smell, the stench was like meat gone bad.

The policeman that stood over Carlos Terrano never saw anything like it and later told his story to fellow police officers, many times; but the story was told only over beers, and usually, after having had several. What he saw was a young kid running, zigzagging through the alley with a big gun in his hand. And seeing that he could not close the gap on the young kid, with the perfect running technique, and seeing that the kid would reach the opening of the alley, where he was sure to lose the kid in the projects that the alley opened into, he stopped, gave the order to "halt or I'll fire," and when the kid did not stop, but rather turned and began to lift the gun, he took aim and fired.

The boy, the officer would say, propelled by the bullet sailed through the air headfirst, landing in the garbage and then immediately began to swim through it. When he caught up to the boy, the right side of his head was gone, and the white-yellowish brain matter oozed from his head and mixed with the garbage. And the kid, although the motion of swimming had long ceased to move him forward, continued to swim through the garbage. "The kid was dead before he hit the ground," the police officer would always say at the end of his story, then add, "but the kid never knew it."

Later that evening, a mother, accompanied by a little girl in a pink dress, bought a box of cereal and a gallon of milk at the Sugar Brown store. When the mother received her change, she counted the coins in her palm, then asked for a small package of balloons behind the counter. This brought joy to the little girl, who began to pull on her mother's yellow sun dress and who in her jubilation stepped on her mother's white sandals. The young mother smiled and stroked the little girl's head to calm her and then placed the coins on the Formica countertop, where the flecks of gold had been rubbed out, and thanked the man.

About the Author

EMMANUEL BURGIN'S non-fiction has appeared in the *Los Angeles Times*, *El Sol de San Diego*, *San Diego Weekly Reader*, *Times of San Diego*, *Rugby West Magazine* and *Rugby News*, His fiction has appeared in *San Diego Writer's Monthly*, *Tidepools*, *Weavings: An Anthology of San Diego Writers*, and *Solstice Quarterly*.

As Sports Editor of *El Sol de San Diego*, Emmanuel covered San Diego's professional football, baseball and soccer teams, and major boxing events held in San Diego, Palm Springs, and Las Vegas. He also reported on education, environment, health, and politics.

Emmanuel was a captain of the Twin City Cougars, a minor-league professional football team, in 1981. Before joining the Cougars, he was under contract with the L.A. Thunderbolts. In 1984, he played a role in establishing the country's largest peer group security company, and in 1986, he served as the Security Director for the Bob Dylan/Tom Petty and the Heartbreakers North American tour. From 1999 to 2020, he was the General Manager of Con Pane Rustic Breads & Café in San Diego, California, which he helped establish.

Despite his various undertakings, Emmanuel has remained dedicated to writing.

Emmanuel, originally from Lynwood, California, pursued his college education at California State University, Northridge, thanks to a football scholarship. Emmanuel lives in Prague, Czechia. He is a member of PEN America.

ALSO BY

Emmanuel Burgin

San Diego Drag Racing and the Bean Bandits

Bean Bandits Four Days at the World of Speed

Vagabond Blues: A Robbie Santos Novel

REQUEST

SUBSCRIBE